ICMI's Pocket Guide to Call Center Management Terms

A convenient, portable reference of terms from ICMI's Call Center Management Dictionary: The Essential Reference for Contact Center, Help Desk and Customer Care Professionals

Brad Cleveland

Published by:
Call Center Press®
A Division of ICMI, Inc.
P.O. Box 6177
Annapolis, Maryland 21401

Printed in the United States of America

ISBN 1-932558-00-4

ICMI's Pocket Guide to Call Center Management Terms

A convenient, portable reference of terms from ICMI's Call Center Management Dictionary: The Essential Reference for Contact Center, Help Desk and Customer Care Professionals

Brad Cleveland

Call Center Press™
A Division of ICMI, Inc.

Contents

Foreword

Everything in the call center industry seems to be expanding—contact channels, customer and employee expectations, technology options and the overall role of the call center. With so much growth and expansion, we, at ICMI, thought it might be refreshing to introduce something small enough to fit inside a busy manager's pocket.

But don't let the size fool you. While *ICMI's Pocket Guide to Call Center Management Terms* may be diminutive in nature, it can have a big impact on the knowledge, eloquence and insight of those working in this dynamic profession.

With a comprehensive—though compact—guide to just about every industry acronym, as well as concise definitions of every term a manager or supervisor should know, *The Pocket Guide* aims to promote consistency and clarity in the way that call center professionals worldwide communicate, cooperate and strive to understand this exciting field.

The ICMI Team

Acronyms and Abbreviations

ACD Automatic Call Distributor

ACS Automatic Call Sequencer

ACW After-Call Work

AHT Average Handling Time

AHT Average Holding Time on Trunks

ANI Automatic Number Identification

ARS Automatic Route Selection

ARU Audio Response Unit

ASA Average Speed of Answer

ASP Application Service Provider

ASR Automatic Speech Recognition

ATA Average Time to Abandonment

ATB All Trunks Busy

AWT Average Work Time

BOC Bell Operating Company

BRI Basic Rate Interface

CBT Computer-Based Training

CCR Customer-Controlled Routing

CCS Centum Call Seconds

CED Caller-Entered Digits

Acronyms and Abbreviations

CIM Customer Interaction Management

CIO Chief Information Officer

CIS Customer Information System

CLEC Competitive Local Exchange Carrier

CLI Calling Line Identity

CMS Call Management System

CO Central Office

CPE Customer Premises Equipment

CRM Customer Relationship Management

CSR Customer Service Representative

CTD Cumulative Trauma Disorder

CTI Computer Telephony Integration

DDD Direct Distance Dialing

DID Direct Inward Dialing

DN Dialed Number

DNIS Dialed Number Identification Service

DSL Digital Subscriber Line

DTMF Dual-Tone Multifrequency

ERMS Email Response Management System

Acronyms and Abbreviations

ERP Enterprise Resource Planning

EWT Expected Wait Time

FCC Federal Communications Commission

FCR First-Call Resolution

FTE Full-Time Equivalent

FX Foreign Exchange Line

GOS Grade of Service

GUI Graphical User Interface

HTML Hyper-Text Markup Language

HTTP Hyper-Text Transport Protocol

ILEC Incumbent Local Exchange Carrier

IM Instant Messaging

IP Internet Protocol

IRR Internal Rate of Return

IS Information Systems

ISDN Integrated Services Digital Network

ISP Internet Service Provider

IT Information Technology

IVR Interactive Voice Response

Acronyms and Abbreviations

IWR Interactive Web Response

IXC Inter-Exchange Carrier

KB Knowledge Base

KPI Key Performance Indicator

LAN Local Area Network

LCD Liquid Crystal Display

LEC Local Exchange Carrier

LED Light-Emitting Diode

LWOP Leave Without Pay

MAC Moves, Adds and Changes

MIS Management Information System

MTBF Mean Time Between Failure

NCC Network Control Center

NIC Network Interface Card

NPA Numbering Plan Area

NPV Net Present Value

OCR Optical Character Recognition

OJT On-the-Job Training

PABX Private Automatic Branch Exchange

Acronyms and Abbreviations

PBX Private Branch Exchange

PC Personal Computer

PCP Post-Call Processing

PDA Personal Digital Assistant

PRI Primary Rate Interface

PSN Public Switched Network

PUC Public Utility Commission

RAN Recorded Announcement

RBOC Regional Bell Operating Company

RFI Request for Information

RFP Request for Proposal

RFQ Request for Quote

ROA Return on Assets

ROI Return on Investment

ROS Return on Sales

RSF Rostered Staff Factor

SBR Skills-Based Routing

SLA Service Level Agreement

SS7 Signaling System 7

Acronyms and Abbreviations

TBT Technology-Based Training

TCP/IP Transmission Control Protocol/Internet Protocol

TSF Telephone Service Factor

TSR Telephone Sales or Service Representative

TTS Text-to-Speech

UCD Uniform Call Distributor

URL Uniform Resource Locator

VDT Video Display Terminal

VOIP/VoIP Voice over Internet Protocol

VPN Virtual Private Network

VRU Voice Response Unit

VXML Voice Extensible Markup Language

WAN Wide Area Network

WAP Wireless Application Protocol

WATS Wide Area Telecommunications Service

WFMS Workforce Management System

WWW World Wide Web

XML Extensible Markup Language

Numbers

0345 Number. A telecommunications service in the United Kingdom; the caller is charged local rates and the organization pays additional required charges (installation, rental and usage/distance charges).

0800 Number. The equivalent of a North American 800 number, used in many countries outside North America for toll-free access.

24x7. Refers to operations that are always open for business (24 hours a day, seven days a week). Often pronounced "twenty-four seven."

2500 Set. A basic touchtone telephone set. See Dual-Tone Multifrequency.

800 Portability. See Number Portability.

802.11b. A set of standards for wireless services.

900 Service. A pay-per-call service where the caller pays a premium charge for the service (e.g., to reach technical support, entertainment services, weather or information lines). 900 is strictly regulated in most countries.

Abandoned Call (Inbound). Also called a lost call. The caller hangs up before reaching an agent. Abandoned calls are available directly from ACD reports. Abandonment rate, though often a primary objective, is not a concrete measure of call center performance because it is driven by caller behavior, which the center cannot directly control; it should be of secondary importance to service level. Related terms: Abandoned Rate (Outbound), Caller Tolerance, Service Level.

Abandoned Rate (Inbound). See Abandoned Call.

Abandoned Rate (Outbound). In a predictive dialing mode, this is the percentage of calls connected to a live person that are never delivered to an agent. Related terms: Abandoned Call (Inbound), Caller Tolerance, Dialer.

Account Code. A code assigned to a specific project, division or client. When making an outbound call, account codes can tie costs back to specific projects.

Activity Codes. See Wrap-Up Codes.

Adherence to Schedule. A general term that refers to how well agents adhere to their schedules. The measure is independent of whether the call center actually has the staff necessary to achieve a targeted service level and/or response time; it is simply a comparison of how closely agents adhere to schedules.

The two terms most often associated with adherence include:

• Availability—The amount of time agents were available.

• Compliance—When they were available to take calls.

Related terms: Occupancy, Real-Time Adherence Software.

After-Call Work (ACW). Also called wrap-up, post call processing, average work time or not ready. Work that is necessitated by and immediately follows an inbound call. Related terms: Average Handling Time, Queue Dynamics, Talk Time.

Agent. The person who handles incoming or outgoing contacts. Also referred to as customer service representative (CSR), customer care representative, telephone sales or service representative (TSR), rep, associate, consultant, engineer, operator, technician, account executive, team member, customer service professional, staff member, attendant or specialist.

Agent Features. Features on a system that are designed for agent use. (Vanguard)

Agent Group. Also called split, gate, queue or skills

group. An agent group shares a common set of skills and knowledge, handles a specified mix of contacts (e.g., service, sales, billing or technical support) and can be comprised of hundreds of agents across multiple sites. Supervisory groups and teams are often subsets of agent groups. Related terms: Cross-Train, Pooling Principle, Queue Dynamics, Skills-Based Routing.

Agent Out Call. An outbound call placed by an agent.

Agent Performance Report. An ACD report that provides statistics for individual agents (e.g., on talk time, after-call work and unavailable time).

Agent Selection. A function in routing software that selects the best agent to handle a call when there is no queue. See Call Selection. (Vanguard)

Agent Status. The mode an agent is in (e.g., talk time, after-call work, unavailable, etc.). See Work State.

Agents. See Average Number of Agents.

All Trunks Busy (ATB). When all trunks are busy in a specified trunk group. Generally, ATB reports indicate how many times all trunks were busy (how many times the last trunk available was seized), and how much total time all trunks were busy. What they don't reveal is how many callers got busy signals when all trunks were busy. Related Terms: Erlang B, Trunk Load.

Analog. Telephone transmission or switching that is not digital. Signals are analogous to the original signal.

Analytics. A general term for advanced reporting and data analysis. In call centers, a set of products or methods that typically interact with CRM systems and multisource data warehouses to collect, analyze and report on particular customer trends or buying patterns. (Vanguard)

Announcement. A recorded verbal message played to callers. See Delay Announcement.

Answer Supervision. The signal sent by the ACD or other device to the local or long-distance carrier to accept a call. This is when billing for either the caller or the call center will begin, if long-distance charges apply.

Answered Call. When referring to an agent group, a call is counted as answered when it reaches an agent. Related terms: Handled Call, Offered Call, Received Call.

Application-Based Routing and Reporting. An ACD capability that enables the system to route and track transactions by type of call, or application (e.g., sales, service, etc.) versus the traditional method of routing and tracking by trunk group and agent group.

Application Service Provider (ASP). An outsourcing business that hosts software applications at its own facilities. Customers "rent" the applications, usually for a monthly fee. Applications are usually accessed via the Internet. (Vanguard)

Architecture. The basic design of a system. Determines how the components work together, system capacity, ability to upgrade and the ability to integrate with other systems.

Attendant. A person who works at a company switchboard, often called a receptionist or operator. See Agent.

Audio Response Unit (ARU). See Interactive Voice Response.

Audiotex. A voice processing capability that enables callers to automatically access pre-recorded announcements. Related terms: Interactive Voice Response, Voice Processing.

A

Auto Available. An ACD feature whereby the ACD is programmed to automatically put agents into available after they finish talk time and disconnect calls. If they need to go into after-call work, they have to manually put themselves there. Related terms: Auto Wrap-Up, Manual Available.

Auto Wrap-Up. An ACD feature whereby the ACD is programmed to automatically put agents into after-call work after they finish talk time and disconnect calls. When they have completed any after-call work required, they put themselves back into available. See Auto Available.

Automated Attendant. A voice processing capability that automates the attendant (operator or receptionist) function. The system prompts callers to respond to choices (e.g., press one for this, two for that...) and then coordinates with the ACD to send callers to specific destinations. This function can reside in an on-site system or in the network. See Interactive Voice Response.

Automated Greeting. An agent's pre-recorded greeting that plays automatically when a call arrives at his or her telephone station.

Automated Reply. A system-generated email that is automatically sent to a customer acknowledging that his or her email was received. Many automated replies also inform the customer of when to expect a response. See Response Time.

Automatic Answer. See Call Forcing.

Automatic Call Distributor (ACD). The specialized telephone system—or more specifically, a software application—that is used in incoming call centers. Basic ACD capabilities include: route calls; sequence calls; queue calls; encourage callers to wait (by playing delay announcements and, in some cases, predicting and announcing wait times); distribute calls among agents; capture planning and performance data, both real-time and historical; and integrate with other systems (the ACD has become just one of many systems in a comprehensive solution).

There are many types of ACDs, including: PBX-based ACD (the ACD is a function on a PBX system); standalone ACD (ACD is the sole function); hybrid ACD (via CTI or an add-on server); key systems; centrex (Central Office-based ACD); third-party managed/hosted ACD Services; and IP Telephony (IP infrastructure with ACD functionality). Related terms: Agent Group, Conditional Routing, Pooling Principle, Skills-Based Routing.

A

Automatic Call Sequencer (ACS). A simple system that is less sophisticated than an ACD, but provides some ACD-like functionality. See Automatic Call Distributor.

Automatic Number Identification (ANI). A telephone network feature that passes the number of the phone the caller is using to the call center in real-time. ANI is an American term; Calling Line Identity (CLI) is an alternative term used elsewhere. Related terms: Computer Telephony Integration, Dialed Number Identification Service.

Auxiliary Work State. An agent work state that is typically not associated with handling telephone calls. When agents are in an auxiliary mode, they will not receive inbound calls.

Availability. The time agents spend handling calls or waiting for calls to arrive. See Adherence to Schedule.

Available State. The work state of agents who are signed on to the ACD and are waiting for calls to arrive. See Occupancy.

Available Time. The total time that an agent or agent group waits for calls to arrive, for a given time period.

Average Call Value. A measure common in revenue-producing call centers. It is total revenue divided by total number of calls for a given period of time.

Average Delay. See Average Speed of Answer.

Average Delay of Delayed Calls. See Average Time to Abandonment.

Average Delay to Abandon. See Average Time to Abandonment.

Average Handling Time (AHT). The sum of average talk time plus average after-call work. Data on AHT is available from ACD reports for incoming calls, and from ERMS and Web servers for email and Web contacts. AHT may also be available from a workforce management system. Related terms: Talk Time, After-Call Work.

Average Holding Time on Trunks (AHT). The average time inbound transactions occupy the trunks. It is:

(Talk time + delay time)/calls received

Related terms: Erlang B, Trunk Load.

Average Number of Agents. The average number of agents logged into an ACD group for a specified time period.

Average Speed of Answer (ASA). A measure that reflects the average delay of all calls, including those that receive an immediate answer. It is available from the ACD. Also called average delay.

Average Time to Abandonment (ATA). Also called average delay to abandon. The average time that callers wait in queue before abandoning. The calculation considers only the calls that abandon. Related term: Caller Tolerance.

Average Work Time (AWT). See After-Call Work.

Back Office. Business applications and functions that are "behind the scenes" to a customer. Examples include accounting, finance, inventory control, fulfillment, productions and human resources. Back-office applications are often associated with enterprise resource planning systems. Related terms: Enterprise Resource Planning, Front Office.

Backbone. The part of the communications network that carries the most traffic (e.g., the services that connect cities, LANs or call centers).

Bandwidth. The transmission capacity of a communications line.

Barge-In. An ACD feature that allows a supervisor or manager to join or "barge-in" on a call being handled by an agent.

Base Staff. Also called seated agents. The minimum number of agents required to achieve service level and response time objectives for a given period of time. Related term: Rostered Staff Factor.

Basic Rate Interface (BRI). One of two basic levels of ISDN service. A BRI line provides two bearer channels for voice and data and one channel for signaling (commonly expressed as 2B+D). Related terms: Integrated Services Digital Network, Primary Rate Interface.

B

Beep Tone. An audible notification that a call has arrived. Beep tone can also refer to the audible notification that a call is being monitored. Also called zip tone. Related terms: Automated Greeting, Call Forcing.

Bell Operating Company (BOC). See Regional Bell Operating Company.

Benchmarking. Historically, a term referred to as a standardized task to test the capabilities of devices against each other. In quality terms, benchmarking is comparing products, services and processes with those of other organizations to identify new ideas and improvement opportunities.

Billing Increment. The unit of time a telecommunications company uses for billing service (e.g., six seconds or 1/10th of a minute).

Binary. A number system based on 1s and 0s. See Digital.

Binaural Headset. A headset with two earpieces, one for each ear. See Headset.

Blended Agent. An agent who handles both inbound and outbound calls, or who handles contacts from different channels (e.g., email and phone). See Call Blending.

Blockage. Callers blocked from entering a queue. See Blocked Call.

Blocked Call. A call that cannot be connected immediately because: A) no circuit is available at the time the call arrives, or B) the ACD is programmed to block calls from entering the queue when the queue backs up beyond a defined threshold. Consequently, data on blocked calls may come from your ACD, local telephone company or long-distance provider. When a call is blocked, the caller hears a busy signal. See Controlled Busies.

Business Rules. A phrase used to refer to various software (or manual) controls that manage contact routing, handling and follow up. Often used interchangeably with workflow. Viewed simplistically, business rules are nothing but a sequence of "if-then" statements. But when the entire range of possible actions that can be undertaken by an organization is represented by such simple statements, the result can be a dense and complex forest of decision trees. See Customer Relationship Management.

Business to Business (B-to-B). Refers to business or interactions between businesses. See Business to Consumer.

Business to Consumer (B-to-C). Refers to business or

interactions between a business and consumers. See Business to Business.

Busy. In use, or "off hook."

Busy Hour. A telephone traffic engineering term, referring to the hour of time in which a trunk group carries the most traffic during the day.

Busy Season. The busiest time of a year for a call center.

Calibration. In a call center, calibration is the process in which variations in the way performance criteria are interpreted from person to person are minimized. In a typical calibration session, the people who routinely monitor agents individually monitor the same call. The ratings and/or scores are then discussed until the group comes to consensus on the most appropriate ratings and/or scores. See Monitoring.

Call. Also called contact or transaction. Although it most often refers to a telephone call, call can also refer to a video call, a Web call and other types of customer contacts. See Call Center.

Call Blending. Traditionally, the ability to dynamically allocate call center agents to both inbound and outbound calling based on conditions in the call center and programmed parameters. This enables a single agent to handle both inbound and outbound calls from the same position without manually monitoring call activity and reassigning the position. The outbound dialing application monitors inbound calling activity and assigns outbound agents to handle inbound calls as inbound volume increases, and assigns inbound agents to outbound calling when the inbound volume drops off. More recently, call blending has evolved to also refer to blending calls with non-phone work or handling contacts from different channels (e.g., email and phone). See Blended Agent.

Call-by-Call Routing. The process of routing each call to the optimum destination according to real-time conditions. Related terms: Network, Network Interflow, Percent Allocation.

Call Center. ICMI defines call center as: "A coordinated system of people, processes, technologies and strategies that provides access to organizational resources through appropriate channels of communication to enable interactions that create value for the customer and organization."

Essentially, call center has evolved into an umbrella term that generally refers to groups of agents handling reservations, help desks, order functions, information lines or customer services, regardless of how they are organized or what types of transactions they handle. Characteristics of a call center generally include:

• Calls (contacts) go to a group of people, not a specific person. In other words, agents are cross-trained to handle a variety of contacts.

• Routing and distribution systems (e.g., ACD systems and/or email response management systems) are generally used to distribute contacts among agents, put calls in queue when all agents are occupied and provide essential management reports.

• Call centers often use advanced network services (e.g., 800 service, DNIS, ANI), and many use interactive voice response capabilities.

• Agents have real-time access to current information via specialized database programs (e.g., status of customer accounts, products, services and other information).

• Management challenges include forecasting calls, calculating staffing requirements, organizing sensible schedules, managing the environment in real-time and getting the right people in the right places at the right times, doing the right things.

As organizations everywhere transition telephone-centric centers into multichannel environments, many are questioning the term call center. Examples of alternative terms include: contact center, interaction center, customer care center, customer support center, customer communications center, customer services center, sales and service center, technical support center, and help desk. Additionally, industry-related terms, such as reservations center (the travel industry), hotline (emergency services) and trading desk (financial services), are commonly found in specific types of organizations.

C

The jury is still out on which terms will emerge as front-runners in coming years. (Note, this guide uses the term call center to refer to any customer contact/customer interaction environment.) See Call Center Value Proposition.

Call Center Initiated Assistance. Typically, this refers to a text-chat session initiated by the agent, rather than the customer.

Call Center Management. ICMI defines call center management as: "The art of having the right number of skilled people and supporting resources in place at the right times to handle an accurately forecasted workload, at service level and with quality."

This definition can be boiled down to two major objectives: 1) Get the right people and supporting resources in the right places at the right times, and 2) do the right things. In other words, provide accessibility with quality. These themes run throughout call center management, from strategic decisions down to day-to-day tactics. See Call Center.

Call Center Value Proposition. The set of benefits that the call center provides to the organization. The following are generally recognized areas in which call centers add strategic value: business unit strategies (call centers are collecting and analyzing more and more information useful to other business units, enabling them to make better strategic and operational decisions); customer satisfaction and loyalty; quality and innovation (call centers provide the organization with customer data to make innovative improvements that will meet the needs of their most valuable customers); marketing; products and services (the call center provides insight into which products and services will best meet the needs of each customer segment); efficient service delivery; self-service usage and system design (call centers provide both the data for determining the best way to offer those services, as well as personal support when self-service options become insufficient); and revenue/sales. Call center managers need to be aware of all the benefits a call center can provide, and need to establish which sets of benefits best fit their call center's role in their organization. See Call Center.

Call Control Variables. The set of criteria the ACD uses to process calls. Examples include routing criteria, overflow parameters, recorded announcements and timing thresholds.

Call Detail Recording (CDR). A telephone system feature that allows the system to record the details of incoming and outgoing calls (e.g., when they occur, how long they last and which extensions they go to). Also called station message detail recording.

Call Forcing. An ACD feature that automatically delivers calls to agents who are available and ready to take calls. They hear a notification that the call has arrived (e.g., a beep tone), but do not have to press a button to answer the call. Sometimes called automatic answer. See Manual Answer.

Call Load. Also called workload. Call load is volume multiplied by average handling time, for a given period of time.

Call Management System (CMS). Another term for an ACD reporting system.

Call Quality (Contact Quality). Typically, a measure that assigns a value to the quality of individual contacts. The following are components of a quality call when viewed at an organizationwide level:

• Customer does not get a busy signal (when using telephone) or "no response" (from Web site)
• Customer is not placed in queue for too long
• Agent provides correct response
• All data entry is correct
• Agent captures all needed/useful information

- Agent has "pride in workmanship"
- Contact is necessary in the first place
- Customer receives correct information
- Customer has confidence contact was effective
- Customer doesn't feel it necessary to check-up, verify or repeat
- People "down the line" can correctly interpret the order
- Customer is not transferred around
- Customer doesn't get rushed
- Customer is satisfied
- Unsolicited marketplace feedback is detected and documented
- Call center's mission is accomplished

Related terms: Monitoring, System of Causes.

Call Recording. A type of monitoring in which the supervisor or automated system records a sampling of calls. The person conducting the monitoring then randomly selects calls for evaluation of agent performance. See Quality Monitoring System.

Call Selection. A function in routing software that selects the best call for an agent to handle when there is a queue and an agent has come available. See Agent Selection. (Vanguard)

Call Treatment. A term that refers generally to

announcements, music, busy signals, ringing or
recorded information provided to callers while they
are in queue. (Vanguard)

Callback Messaging. A feature that enables callers
waiting in queue to leave a message or to enter
their telephone numbers for later callback from an
agent.

Caller Entered Digits (CED). The digits a caller enters
on his or her telephone keypad. Usually used for
auto attendant, voice response and CTI applications.
Also referred to as prompted digits.

Caller ID. See Automatic Number Identification.

Caller Tolerance. How patient callers will be when
they encounter queues or busy signals. There are
seven factors of caller tolerance, which include:

1. Degree of motivation
2. Availability of substitutes
3. Competition's service level
4. Level of expectations
5. Time available
6. Who's paying for the call
7. Human behavior

These factors influence such things as how long
callers will wait in queue, how many callers will
abandon, how many will retry when they get busy

signals, how they will react to automation, such as IVR or Web services, and how they perceive the service the call center is providing. Related terms: Abandoned Call, Delay Announcements.

Calling Line Identity (CLI). See Automatic Number Identification.

Calls in Queue. A real-time report that refers to the number of calls received by the ACD system but not yet connected to an agent.

Career Path. Career paths guide individual employee development through structured advancement opportunities within the call center and/or organization. Most career paths require specific tasks to be successfully accomplished in order for an employee to move from one level to the next. A typical career path model requires the development of job families, which are comprised of a number of jobs arranged in a hierarchy by grade, pay and responsibility (e.g., agent, team leader, supervisor, manager, senior manager and director). See Skill Path.

Carrier. A company that provides telecommunications circuits. Carriers include both local telephone companies, also called local exchange carriers (LECs), and long-distance providers, also called inter-exchange carriers (IXCs).

C

Cause-and-Effect Diagram. A chart that illustrates the relationships between causes and a specific effect you want to study.

Central Office (CO). Can refer to either a telephone company switching center or the type of telephone switch used in a telephone company switching center. The local central office receives calls from within the local area and either routes them locally or passes them to an inter-exchange carrier (IXC). On the receiving end, the local central office receives calls that originated in other areas from the IXC.

Centrex. A central office telephone switch service that serves a specific area. (Vanguard)

Centum Call Seconds (CCS). A unit of telephone traffic measurement referring to 100 call seconds. The first C represents the Roman numeral for 100. 1 hour of telephone traffic=1 Erlang=60 minutes=36 CCS. Related terms: Erlang, Erlang B, Erlang C.

Chief Information Officer (CIO). A typical title for the highest ranking executive responsible for an organization's information systems.

Circuit. A transmission path between two points in a network.

Circuit Switching. A method of transferring information across a network by establishing a temporary, dedicated, end-to-end path (a circuit) for the duration of a communication. This is the technology traditionally used to transmit voice (e.g., over the public switched telephone network). (Vanguard)

Client/Server Architecture. A networked computing approach in which one computer application (client) issues a request to another computer application (server). The server application processes the request and delivers the requested information back to the client application. Related term: Desktop Technologies. (Vanguard)

Co-Browsing. A term that refers to the capability of both an agent and customer to see a Web page simultaneously and share navigation and data entry.

Collateral Duties. Non-phone tasks (e.g., data entry) that are flexible and can be scheduled for periods when call load is slow. Related terms: Schedule, Schedule Alternatives.

Communications Server. An alternative to the PBX that manages and routes voice, fax, Web and email communications within a single server and provides a wide set of applications. (Vanguard)

Competitive Local Exchange Carrier (CLEC). See Local Exchange Carrier.

C

Completed Call. A general term that refers to an inbound contact that successfully reaches and is handled by an agent. Can also refer to an outbound call that successfully reaches a live person (or answering machine, if leaving a message is acceptable). In an outbound context, also called connected call.

Compliance. See Adherence to Schedule.

Computer-Based Training (CBT). Training programs delivered through software applications without the need for a facilitator.

Computer Simulation. A computer-based simulator program that predicts the outcome of various events in the future, given many variables. In call centers, it is most often used to determine staff required to meet service levels and response times in complex routing environments.

Computer Telephony Integration (CTI). CTI integrates the functions of telephone networks, voice switching, data switching, computer applications, databases, voice processing and alternative media. With this comes the ability to exchange commands and messages between systems. This results in the ability to monitor and control calls, events, applications, information and endpoints. CTI can add or enhance functionality in a number of areas: coordinating voice and data, intelligent routing, integrated reporting, desktop softphone and outbound dialing. (Vanguard)

Concentrated Shift. A scheduling technique that requires agents to work more hours in a day, but fewer days in a week. "Four-by-10" shifts (four days on for 10 hours each, with three days off) are particularly popular with many agents.

Conditional Routing. The capability of the ACD to route calls based on real-time criteria (e.g., calls in queue, time of day and type of call). It is based on "if-then" programming statements. For example, "if the number of calls in agent group one exceeds 10 and there are at least two available agents in group two, then route the calls to group two."

Contact History. The history of a customer's interactions with an organization, generally recorded and stored in a customer information system.

Contact Management System. Business application that enables and tracks each interaction with the customer. (Vanguard)

Contacts Handled (Calls Per Agent). The number of contacts an agent handles in a given period of time. Related terms: Occupancy, Queue Dynamics, True

C

Calls Per Agent.

Contacts Per Hour. An outbound term that refers to the number of contacts divided by agent hours on the dialer. See Contacts.

Continuous Improvement. The ongoing improvement of processes. See System of Causes.

Control Chart. A quality tool that provides information on variation in a process. Quality problems in the call center are challenging and often confusing because they are a part of a complex process, and any process has variation from the ideal. There are two major types of variation: special causes and common causes. Special causes create erratic, unpredictable variation. Common causes are the rhythmic, normal variations in the system. A control chart enables you to bring a process under statistical control by eliminating the chaos of special causes. You can then work on the common causes by improving the system and, thus, the whole process.

Controlled Busies. The capability of the ACD to generate busy signals when the queue backs up beyond a programmable threshold. See Blocked Call.

Conventional Shift. A traditional five-day-a-week shift during "normal business hours" (e.g., 9 a.m. to

5 p.m., Monday through Friday).

Cookie. A small file in a Web browser that uniquely identifies a user to a Web server to provide personalized content. (Vanguard)

Coordinated Voice/Data Transfer. A CTI application similar to coordinated voice/data conference, except that the voice call and the data are transferred to a colleague. Often used when transferring a call from an IVR to an agent position with a screen pop. (Vanguard)

Cost/Benefit Analysis. A term used to describe the process of comparing the value of a potential project with the cost associated with it.

Cost Center. An accounting term that refers to a department or function in the organization that does not generate profit. When a call center is viewed as a cost center, the focus is on getting the transactions done at the least total cost to the organization. Related term: Profit Center.

Cost of Delay. The direct expense of putting callers in queue. In an inbound call center, each person connected to your system requires a trunk, whether they are talking to an agent or waiting in queue. If you have toll-free service (or any other service that charges a usage fee), you are paying for this time.

C

The cost of delay is expressed in terms of how much you pay for your network service each day for callers to wait in queue until they reach an agent.

Cost Per Call. Total costs (fixed and variable) divided by total calls for a given period of time.

Cradle-to-Grave Reporting. A call center reporting term that includes all call center touch points (human and systems) from the time a caller dials an 800 number to the time of disconnect. It can include, but is not limited to, voice-switch routing, IVR, multisite flows, all agent activity and business application activity. The key enabler for cradle-to-grave reporting is typically CTI. (Vanguard)

Cross-Sell. A suggestive selling technique that offers additional products or services to current customers, usually based on relationships established between the customer's profile and the attributes of customers who have already purchased the products or services being cross-sold.

Cross-Train. To train agents to handle more than one defined mix of calls (e.g., to train technical support agents handling laptop calls to also handle desktop issues).

Cumulative Trauma Disorder (CTD). See Ergonomics.

Customer Access Strategy. In the call center environment, strategy is embodied in what is often termed a "customer access strategy," which is a framework—a set of standards, guidelines and processes—defining the means by which customers are connected with resources capable of delivering the desired information and services.

Customer Care. A general term that refers to proactive customer service that creates high levels of customer satisfaction and loyalty. The term customer care center has become an alternative to call center, particularly in some countries outside North America. See Call Center.

Customer Contact. See Call.

Customer Controlled Routing (CCR). A vendor-specific term (originated by Nortel) that refers to a call routing application that enables calls to be handled (e.g., routed, queued, distributed) based on user-defined criteria.

Customer Expectations. The expectations customers have of a product, service or organization. When interacting with organizations, ICMI studies have shown that customers have 10 primary expectations, which include (in no specific order):
    1. Be accessible

2. Treat me courteously
3. Be responsive to what I need and want
4. Do what I ask promptly
5. Provide well-trained and informed employees
6. Tell me what to expect
7. Meet your commitments; keep your promises
8. Do it right the first time
9. Be socially responsible and ethical
10. Follow up

Related terms: Call Quality, Service Level, Response Time.

Customer Information System (CIS). Also called customer interaction software. A database application (or series of integrated applications) that provides information about the customer (e.g., their interactions with the organization, the services and products they have purchased). (Vanguard)

Customer Interaction Management (CIM). See Customer Relationship Management.

Customer Lifetime Value. Expresses the value of a customer to the organization over the entire probable time period that the customer will interact with the organization. Related terms: Customer Loyalty, Customer Relationship Management.

Customer Loyalty. Typically defined in terms of the

customer's repurchase behavior, intent to purchase again or intent to recommend the organization. Related terms: Customer Relationship Management, Customer Satisfaction.

Customer Premises Equipment (CPE). A telecommunications term referring to equipment installed on the customer's premises and connected to the telecommunications network.

Customer Profiling. The process of collecting and maintaining information about customers and their relationship to your organization. The information is most often used for customer segmentation and building customer relationships.

Customer Relationship Management (CRM). The process of holistically developing the customer's relationship with the organization. It takes into account their history as a customer, the depth and breadth of their business with the organization, as well as other factors. Customer relationship management generally uses applications and database systems that include elements of data mining, contact management and enterprise resource planning, enabling agents and analysts to know and anticipate customer behavior. Related terms: Business Rules, Customer Loyalty, Customer Satisfaction, Customer

C

Retention Rate.

Customer Retention Rate. The percentage of a prior period's customers who are still customers in the current period (excluding new customers acquired). Related terms: Customer Loyalty, Customer Relationship Management.

Customer Satisfaction. The level of satisfaction customers have with the organization and the organization's products and services.

Customer Segmentation. The process of grouping customers based on what you know about them, in order to apply differentiated marketing, relationship and contact treatment strategies.

Customer Service Representative (CSR). See Agent.

Customer Survey. A process that acquires and assesses feedback from customers or prospective customers.

Customer Valuation Model. A mathematical formula that estimates the value of a customer to the organization over a future time period. Value can represent both the tangible and intangible benefits of the customer relationship. See Customer Lifetime Value.

Cutover. The date and time that a new system is put into use.

Data-Directed Routing. A routing capability that uses a database of information about the customer, current status or other factors to make routing decisions. Generally it is CTI-enabled. (Vanguard)

Data Mining. The use of analytical tools to uncover correlation between disparate sets of data. The ultimate objective is to understand and serve customers better. (Vanguard)

Data Warehousing. A large database that stores data generated by an organization's multiple business systems. Related term: Data Mining. (Vanguard)

Database. An application containing data that is organized in a structured fashion for quick access.

Database Call Handling. A CTI application whereby the ACD works in sync with the database computer to process calls based on information in the database. For example, a caller inputs his or her account number or other type of identifier into a voice processing system, the database retrieves information on that customer and then issues instructions to the ACD on how to handle the call (e.g., where to route the call, what priority the call should be given in queue, the announcements to play).

Day-of-Week Routing. A network service that routes calls to alternate locations, based on the day of week. There are also options for day-of-year and time-of-day routing.

Delay. Also called queue time. The time a caller spends in queue waiting for an agent to become available. Average delay is the same thing as average speed of answer. Related terms: Average Delay of Delayed Calls, Average Speed of Answer.

Delay Announcement. A recorded announcement designed and positioned to encourage callers to wait for an agent to become available, have information ready (e.g., their account number), provide information pertaining to the delay, or suggest other access alternatives or times callers can get assistance. Related terms: Abandoned Call, Caller Tolerance, Fast Clear Down, Queue.

Dialed Number (DN). The number that the caller dialed to initiate the call.

Dialed Number Identification Service (DNIS). A string of digits that the telephone network passes to the ACD, VRU or other device to indicate which telephone number the caller dialed. The ACD can then process and report on that type of call according to user-defined criteria. One trunk group can have many DNIS numbers. See Automatic Number Identification.

Dialer. Dialers are technologies (hardware/software) for automating the process of making outbound calls to lists of people. In addition to placing outbound calls, dialers may provide campaign management and scripting functionality, track the disposition of calls and provide detailed real-time and historical reporting. Predictive dialing is an application that instructs the switch to dial multiple simultaneous calls from a preloaded list of phone numbers. It seeks to match the number of completed calls with the number of available agents so that completed calls are immediately handled by an agent. Agents also receive a data screen about the call. The system classifies all calls launched (e.g., connect, busy, no answer, answering machine, network tones) and updates the database accordingly. Related terms: Abandoned Rate (Outbound), Completed Call.

Digital. The use of a binary code—1s and 0s—to represent information.

Digital Subscriber Line (DSL). An integrated, digital, high-speed (>384 kbps) Internet access and voice service for small offices and residential users. For call centers, it is an enabling technology for work-at-home (telecommuting) agents. (Vanguard)

Direct Call Processing. See Talk Time.

D

Direct Distance Dialing (DDD). Permits users to place long-distance calls without the assistance of an operator. Related term: North American Numbering Plan.

Direct Inward Dialing (DID). A network service offering—generally associated with local service—in which a unique set of identifying digits is passed to the customer premises equipment. By mapping each set of digits to an internal extension, the switch can provide direct dialing to a particular extension. See Dialed Number Identification Services. (Vanguard)

Directed Dialog. Speech recognition approach that recognizes what is being said based on guided or structured interactions. The caller is given examples of phrases to use. Also referred to as structured language. See Speech Recognition. Related term: Natural Language. (Vanguard)

Disaster Recovery Plan. A plan that enables managers to avoid or recover expediently from an interruption in the center's operation. Comprehensive plans should include an approved set of arrangements and procedures for facilities, networks, people and service levels.

Display Board. See Readerboard.

Distributed Call Center. See Virtual Call Center.

Double Jack. The ability to plug two headsets into one telephone or workstation so that two people can listen to or participate in the same contact. Often used for side-by-side monitoring. Related terms: Monitoring, Headset.

Downtime. The time that a system is unavailable. For example, inbound telephone systems are expected to be operational 99.999 percent of the time, which is commonly referred to as "five nines." (Vanguard)

Driver-Based Forecasting. A form of explanatory forecasting. Any method of workload forecasting that is based on other identified activities or "drivers." See Forecasting Methodologies.

Dual-Tone Multifrequency (DTMF). A signaling system that sends pairs of audio frequencies to represent digits on a telephone keypad. It is often used interchangeably with the term Touchtone (an AT&T trademark).

Dumb Switch. A switch that contains only basic hardware and software, and receives call-handling instructions from another device.

Dynamic Answer. An ACD feature that automatically reconfigures the number of rings before the system answers calls based on real-time queue information.

D

Since costs don't begin until the ACD answers calls, this feature can save callers or the call center money on long-distance charges.

E1 Circuit. See T1 Circuit.

Electronic Board. See Readerboard.

Electronic Mail (Email). The transmission of information in electronic form from one person (or system) to another over the Internet or other computer network.

Email Response Management System (ERMS). An ERMS controls the flow and tracking of email into an organization in much the same way that an ACD controls the flow and tracking of inbound calls. An ERMS can perform the following routing and reporting functions: conditional routing based on skills, customer priority, etc.; priority queuing; response time tracking; and management reporting. (Vanguard)

Employee Satisfaction Survey. An instrument designed to determine which aspects of employees' work situations contribute to or impede their job satisfaction. Typical areas of measurement include management effectiveness, peer/team members effectiveness, career opportunities, equity of salary, level of appreciation, satisfaction with benefits, and quality of training, feedback and coaching.

Enterprise Resource Planning (ERP). A large-scale business application or set of applications that encompass some or all aspects of back-office functions. Related terms: Back Office, Customer Relationship Management, Front Office. (Vanguard)

Envelope Strategy. A scheduling approach whereby enough agents are scheduled for the day or week to handle both the inbound call load and other types of work. Priorities are based on the inbound call load. When call load is heavy, all agents handle calls, but when it is light, some agents are reassigned to work that is not as time-sensitive.

Ergonomics. The science of fitting the job to the worker. Ergonomics can prevent work-related cumulative trauma disorders (CTDs) that result when there is a mismatch between the physical capacity of workers and the physical demands of their jobs.

Erlang. One hour of telephone traffic in an hour of time. For example, if circuits carry 120 minutes of traffic in an hour, that's two Erlangs. Related terms: Erlang B, Erlang C, A.K. Erlang (listed as Erlang, A.K.), Queue Dynamics.

Erlang, A.K. A Danish engineer who worked for the Copenhagen Telephone Company in the early 1900s and developed Erlang B, Erlang C and other telephone traffic engineering formulas. Related terms: Erlang, Erlang B, Erlang C.

Erlang B. A formula widely used to determine the number of trunks required to handle a known trunk load during a one-hour period.

The formula assumes that if callers get busy signals, they go away forever, never to retry ("lost calls cleared"). Since some callers retry, Erlang B can underestimate trunks required. However, Erlang B is generally accurate in situations with few busy signals. Related terms: Erlang, Erlang C, Queue Dynamics, Trunk Load.

Erlang C. A mathematical tool used to calculate predicted waiting times (delay) based on three things: the number of servers (agents); the number of people waiting to be served (callers); and the average amount of time it takes to serve each person. It can also predict the resources required to keep waiting times within targeted limits. Erlang C assumes no lost calls or busy signals, so it has a tendency to overestimate staff required.

Erlang C is widely used in workforce management software programs, as well as low-cost call center staffing calculators. Related terms: Computer Simulation, Erlang, Erlang B, Queue Dynamics.

Error Rate. The number or percentage of defective

(e.g., incomplete) transactions or the number or percentage of defective steps in a transaction.

Errors and Rework. As a measurement, the percent (and types) of errors and rework that are occurring.

Escalation Plan. A plan that specifies actions to be taken when the queue begins to build beyond acceptable levels. See Real-Time Management.

Ethernet. A standard networking technology for putting information on a local area network. (Vanguard)

Event-Driven Forecasting. Any method of workload forecasting that is based on individual activities that generate call volume. See Forecasting Methodologies.

Exchange Line. See Trunk.

Expert System. Also known as a knowledge-based system. A business application that aids the user in analyzing and resolving problems based on logic trees and known solutions to identified problems. Includes functions such as problem analysis and problem resolution. (Vanguard)

Explanatory Forecasting. See Forecasting Methodologies.

Extranet. Networks typically connected via the

Internet, providing for direct and secure business-to-business access between suppliers and vendors or other partners. (Vanguard)

Facsimile (FAX). Technology that scans a document, encodes it, transmits it over a telecommunications circuit, and reproduces it in original form at the receiving end. Related terms: Fax on Demand, Fax Server.

Fast Clear Down. A caller who hangs up immediately after hearing a delay announcement. Related term: Delay Announcement.

Fault Tolerant. The ability for a system or piece of equipment to keep working even if it encounters a hardware failure. Related term: Disaster Recovery Plan.

Fax on Demand. A system that enables callers to request documents, using their telephone keypads. The selected documents are delivered to the fax numbers they specify. Related terms: Facsimile, Fax Server.

Fiber Optics. Thin filaments of transparent glass or plastic that use light to transmit voice, video or data signals.

First-Call Resolution. The percentage of calls that do not require any further contacts to address the customer's reason for calling. The customer does not need to contact the call center again to seek resolution, nor does anyone within the organization need

F

to follow up. Related term: Errors and Rework.

Flex-Time Scheduling. Several weeks in advance, agents are promised schedules within a window of time (e.g., only Tuesdays through Saturdays or from 8 a.m. to 8 p.m. any day of the week), according to their personal availability. Then, specific work hours, and in some cases, days worked, are determined from week to week as forecasted staff requirements are refined. This approach may involve the entire staff, but usually includes only a subset of employees.

Flow Chart. A flow chart is a "map" of a process that is used to analyze and standardize procedures, identify root causes of problems and plan new processes.

Flushing out the Queue. A real-time management term that refers to changing system thresholds so that calls waiting for an agent group are redirected to another group with a shorter queue or more available agents. Related term: Real-Time Management.

Forecasted Call Load vs. Actual. A performance objective that reflects the percent variance between the call load forecasted and the call load actually received. Related term: Forecasting Methodologies.

Forecasting. The process of predicting call center workload and other activities. See Forecasted Call Load vs. Actual and Forecasting Methodologies.

Forecasting Methodologies. General methods used to predict future events, such as the amount of workload that will come into an incoming call center in future time periods. Methodologies are broadly categorized into quantitative and judgmental approaches. Quantitative forecasts include: time-series forecasts, which assume past data will reflect trends that continue into the future; and explanatory forecasting, which essentially attempts to reveal a linkage between two or more variables. Driver-based and event-driven forecasting approaches are variations of explanatory forecasting. Judgmental forecasts go beyond purely statistical techniques. They involve intuition, interdepartmental committees, market research and executive opinion. Related terms: Driver-Based Forecasting, Forecast, Forecasted Call Load vs. Actual.

Foreign Exchange Line (FX). Provides local telephone service from a central office (CO) to a location outside the CO's serving area. In call center applications, FX lines can give dispersed callers local numbers to dial, while contacts come into a central call center.

F

Front Office. Business applications that deal with customer interactions, such as customer service, help desk, sales or customer relationship management. Related terms: Back Office, Enterprise Resource Planning. (Vanguard)

Full-Time Equivalent (FTE). A term used in scheduling and budgeting whereby the number of scheduled hours is divided by the hours in a full work week. The hours of several part-time agents may add up to one FTE. A full work week (e.g., 40 hours per week) represented by one weekly FTE may be contributed by one person working 40 hours, two people working 20 hours, four people working 10 hours, 40 people working one hour, etc.

Gate. See Agent Group.

Gateway. A server dedicated to providing access to a network. Also, software and hardware that interprets and translates different protocols from different networks or devices. (Vanguard)

Grade of Service (GOS). The probability that a call will not be connected to a system because all trunks are busy. Grade of service is often expressed as "p.01" meaning 1 percent of calls will be "blocked." Sometimes, grade of service is used interchangeably with service level, but the two terms have different meanings. Related terms: Erlang B, Service Level, Trunk Load.

Graphical User Interface (GUI). A computer interface that is graphical in nature, and uses menus, icons and a mouse to enable the user to interact with the system. Web browsers and the Windows and Apple operating systems are examples of GUI interfaces.

Handled Call. A call that is received and handled by an agent or peripheral equipment. Related terms: Answered Call, Offered Call, Received Call.

Handling Time. The time an agent spends in talk time and after-call work handling a transaction. Handling time can also refer to the time it takes for a machine to process a transaction. See Average Handling Time.

Headset. A device that consists of an earpiece and a microphone, and replaces a telephone handset. Headsets are designed to fit comfortably on the user's head, freeing both hands.

Heating, Ventilation and Air Conditioning (HVAC). Refers to the systems that control the climate (heating, air conditioning and vitalization) in a building.

Help Desk. A term that generally refers to a call center that provides technical support (e.g., queries about product installation, usage or problems). The term is most often used in the context of computer software and hardware support centers.

Historical Forecasting. Any method of call volume forecasting that relies solely on past call volume to determine future projections. Forecasts from workforce management systems generally rely on historical forecasting.

H

Historical Report. A report that tracks call center and agent performance over a period of time. See Real-Time Report.

Holding Time. See Average Holding Time on Trunks.

Home Agent. See Remote Agent.

Hotline. A call center set up to handle emergency or urgent calls (e.g., an accident hotline).

Hyper-Text Markup Language (HTML). A language derived from the Standard Generalized Markup Language (SGML), primarily used to create Web pages. (Vanguard)

Hyper-Text Transport Protocol (HTTP). A protocol in the Web environment that links addresses with a Web server and presents the appropriate HTML pages. (Vanguard)

Identified Ringing. A telephone system feature that provides distinctive ringing sounds for different types of calls.

Idle Time. The inverse of occupancy. The time agents are available and waiting for contacts to arrive. See Occupancy.

Imaging. A technology to scan printed documents such as mail and transform them into electronic documents for processing, storage and/or routing. (Vanguard)

Immutable Law. A law of nature that is fundamental and not changeable (e.g., the law of gravity). In an inbound call center, the fact that occupancy goes up when service level goes down is an immutable law. See Queue Dynamics.

Incremental Revenue Analysis. A methodology that estimates the value (cost and revenue) of adding or subtracting an agent. This approach determines the potential impact of abandonment because of customer wait time on overall costs; you attach a cost to abandoned calls and make assumptions around how many calls you would lose for various service levels.

Increments. In call centers, increments are the timeframes used for staffing and reporting. Given the variation in workload throughout the day, staff requirements must be calculated at specific increments (which are generally the smallest units of time reflected in the forecast).

Index Factor. In forecasting, a proportion used as a multiplier to adjust another number. For example, in a time-series forecast, Monday's index factor may be 1.2—meaning that Mondays typically receive 1.2 times (20 percent more than) the average day's call load.

Information Systems (IS). A generic term for systems that perform data processing.

Information Technology (IT). A generic term that refers either to computer and/or communications systems and technologies, or the profession that develops and manages these systems.

Instant Messaging (IM). A form of text-chat used primary for non-commercial communications between two or more Internet users. (Vanguard)

Integrated Services Digital Network (ISDN). A set of international standards for telephone transmission. ISDN provides an end-to-end digital network, out-of-band signaling and greater bandwidth than older telephone services. Often used in call centers to deliver signaling information quickly for use of ANI and DNIS, and for faster call setup and tear-down. The two standard levels of ISDN are basic rate interface (BRI) and primary rate interface (PRI). Related terms: Basic Rate Interface, Primary Rate Interface.

Intelligent Routing. The use of information about the caller, current conditions or other parameters to route calls to the appropriate group, individual, automated system, etc. DNIS, ANI, customer-entered digits and database information can all be used as routing parameters. It can augment or replace conditional and skills-based routing performed on the switch and is generally enabled via CTI.

Inter-Exchange Carrier (IXC). A long-distance telephone company.

Interactive Voice Response (IVR). A system that enables callers to use their telephone keypad (or spoken commands if speech recognition is used) to access a company's computer system for the purpose of retrieving or updating information, conducting a business transaction, or routing their call. Also referred to as a voice response unit (VRU). (Vanguard)

Interactive Web Response (IWR). Systems that enable

customers to use the Internet to access a company's Web site for the purpose of retrieving or updating information or conducting a business transaction. (Vanguard)

Interconnect Company. A company that provides telecommunications systems for connection to the telephone network. The term goes back to the days before divestiture; any vendor other than AT&T that provided systems to customers were interconnect companies.

Interflow. See Overflow.

Internal Part-Timers. A scheduling approach, sometimes called the reinforcement method. When contact-handling duties are combined with other types of tasks, such as correspondence, outbound calling or data-entry, the agents assigned to these collateral duties can act as reinforcements when the calling load gets heavy. This is like being able to bring in part-timers on an hourly, half-hourly or even five-minute basis. Related terms: Schedule, Schedule Alternatives.

Internal Response Time. The time it takes an agent group that supports other internal groups (e.g., for complex or escalated tasks) to respond to transactions that do not have to be handled when they arrive. See Response Time.

Internal Turnover. See Turnover.

Internet. A worldwide, expanding network of linked computers, founded by the U.S. government and several universities in 1969, originally called Arpanet and based on TCP/IP protocol. Made available for commercial use in 1992. (Vanguard)

Internet "Call Me" Transaction. A transaction that allows a user to request a callback from the call center, while exploring a Web page. Requires interconnection of the ACD system and the Internet by means of an Internet gateway.

Internet "Call-Through" Transaction. Refers to the ability for callers to click a button on a Web site and be directly connected to an agent (initiate a voice conversation) while viewing the site. Standards and technologies that provide this capability are in development. Also referred to as "click-to-talk."

Internet Phone. Technology that enables users of the Internet's World Wide Web to place voice telephone calls through the Internet, thus by-passing the long-distance network.

Internet Protocol (IP). The set of communication

standards that control communications activity on the Internet. An IP address is assigned to every computer on the Internet.

Internet Service Provider (ISP). A company that provides Internet access to customers, either through a modem or direct connection. Related term: Network Service Provider. (Vanguard)

Interval Based Accuracy. A method of measuring forecast success that focuses on results by interval (usually half-hours), rather than end of day, week or month results. See Forecasting Methodologies.

Intraday Forecast. A short-term forecast that assumes activities early in the day will reflect how the rest of the day will go.

Intraflow. See Overflow.

Intranet. A company's private data network that is accessed using browser-based technology and TCP/IP protocol. (Vanguard)

Intraweek Forecast. A short-term forecast that assumes activities early in the week will reflect how the rest of the week will go.

Invisible Queue. When callers do not know how long the queue is or how fast it is moving. Related terms: Queue, Visible Queue.

IP Phone. An end-user device that enables users to place voice calls through a data network (LAN, WAN or the Internet) using the Internet Protocol. The device can be an IP-enabled telephone or a PC with soundcard and software. Either device converts the source information from circuit-switched to packet-switched format. (Vanguard)

IP Telephony. Technology that enables voice telephone calls to be carried over a data network (a private intranet or the public Internet) using protocols from the TCP/IP suite. Voice is transmitted in data packets. Also referred to as Internet telephony. (Vanguard)

Job Evaluation. The process of identifying and describing jobs, and determining the relative worth or value of a job to the organization.

Job Role. The function or responsibilities related to a specific position in an organization.

Judgmental Forecasting. Goes beyond purely statistical techniques and encompasses what people believe is going to happen. It is in the realm of intuition, interdepartmental committees, market research and executive opinion. See Forecasting Methodologies.

Key Performance Indicator (KPI). A high-level measure of call center performance. Note, some interpret KPI as the single most important measure in a department or unit; however, in common usage, most call centers have multiple KPIs. See Performance Objective.

Knowledge Management. According to consultant Jenny McCune, knowledge management is the task of developing and exploiting an organization's tangible and intangible knowledge resources. The main objective of knowledge management is to leverage and reuse resources that already exist in the organization so people will not spend time "reinventing the wheel."

Knowledge Worker. A worker who is involved primarily in "symbolic analysis"–dealing with symbols, concepts and communications, rather than tangible goods. In this sense, many call center agents are knowledge workers.

Law of Diminishing Returns. The law of diminishing returns, common in economics, is a significant consideration in a queuing environment. It can be stated this way: When successive individual agents are assigned to a given call load, marginal improvements in service level that can be attributed to each additional agent will eventually decline. Related terms: Immutable Law, Queue Dynamics.

Least-Occupied Agent. A method of distributing calls to the agent who has the most idle time (lowest occupancy), in a given period of time. Related terms: Longest-Available Agent, Next-Available Agent. (Vanguard)

Leave Without Pay (LWOP). Agents are offered the chance to leave work early, without pay, when call volumes are low. Pronounced "el-wop." Related terms: Schedule, Schedule Alternatives.

Legacy Systems. Information systems or databases that house vital business information, such as customer records, but are often based on older technologies (e.g., mainframes, mini-computers). (Vanguard)

Liquid Crystal Display (LCD). An alphanumeric display that uses liquid crystal sealed between pieces of glass.

L

Load Balancing. In a network call center environment, load balancing is the process of distributing (balancing) contacts between call centers. Related terms: Call-by-Call Routing, Network, Network Control Center, Network Interflow, Percent Allocation.

Local Area Network (LAN). The connection of multiple computers within a building so that they can share information, applications and peripherals. Related term: Wide Area Network.

Local Exchange Carrier (LEC). A telephone company responsible for providing local connections and services. New startup LECs are sometimes referred to as competitive local exchange carriers (CLECs), while telephone companies in existence at the time of the breakup of AT&T are known as incumbent local exchange carriers (ILECs).

Logged On. A state in which agents have signed on to a system (made their presence known), but may or may not be ready to receive calls.

Logical Agent. An agent identified by their login code, not by their physical position or phone number. This feature enables an agent to login from anywhere in the call center and be recognized by the system the same way for routing and statistics purposes. See Agent Group. (Vanguard)

Long Call. When average handling time approaches or exceeds 30 minutes. Long calls cause problems in call centers that use the typical 30-minute increment for forecasting and staffing. Since long calls are not distributed as Erlang C assumes, they may violate the assumptions of the formula.

Longest-Available Agent. Also referred to as most-idle agent. A method of distributing calls to the agent who has been sitting idle the longest. With a queue, longest available agent becomes next available agent. Related terms: Least-Occupied Agent, Next-Available Agent.

Longest Delay. Also called oldest call. The longest time a caller has waited in queue, before abandoning or reaching an agent. See Queue Dynamics.

Look-Ahead Queuing. The ability for a system or network to examine a secondary queue and evaluate the conditions before overflowing calls from the primary queue.

Look-Back Queuing. The ability for a system or network to look back to the primary queue after the call has been overflowed to a secondary queue and evaluate the conditions. If the congestion clears, the call can be sent back to the initial queue.

L

Lost Call. See Abandoned Call.

Mainframe. A computer system that is a large, monolithic system. It generally has its own operating system, and databases and applications resident on the same system. (Vanguard)

Make Busy. To make a circuit or terminal unavailable.

Managed Staffing Arrangement. An arrangement whereby a managed staffing company supplies all or part of the organization's employee needs, according to the organization's business rules and guidelines. This staffing arrangement is a variation of outsourcing, with the managed staffing company using your facilities instead of their own.

Management Information System (MIS). For call centers, a system that facilitates the capture and reporting of activity within the telephony and computing infrastructure. (Vanguard)

Manual Answer. The ACD system is set up so that agents must manually answer calls. See Call Forcing.

Manual Available. The ACD system is set up so that agents must put themselves back into the available mode after completing any after call work. See Auto Available.

M

Market Research. The disciplined process of collecting, analyzing and interpreting information about customers in order to make better decisions about meeting customer needs and expectations. The call center can both benefit from market research and assist in market research efforts.

Mean Time Between Failure (MTBF). An estimate of the average time before a system or component will likely fail.

Measurement. A quantifiable unit. In call centers, this generally refers to time (e.g., handle time), an input (e.g., a telephone call, email, customer), an output, (e.g., a sale, proposal, completed contact, problem resolution) or a ratio expressed with a numerator and denominator (e.g., absenteeism, close ratio, first-call resolution).

Merlang. A term used by workforce management vendor Pipkins that refers to a modified Erlang formula. Related terms: Erlang, Erlang B, Erlang C.

Metrics. Another word for measurements or, sometimes in usage, objectives. Related terms: Key Performance Indicator, Performance Objective.

Middleware. A generic term for software that mediates between different types of hardware and software on a network so that they can function together. Typically uses open interfaces and applications programming interfaces (APIs) to access and move information. In call centers, middleware is typically used in CRM and CTI application integration. (Vanguard)

Modem. A contraction of the terms modulator/demodulator. A modem converts analog signals to digital, and vice versa.

Modular Jack. An interface that permits easy interconnection of telecommunications equipment or circuits.

Monitoring. Monitoring is a call evaluation process that appraises the qualitative aspects of call handling. Monitoring programs include the tracking and analysis of data to identify individual agent and overall call center performance trends, anticipated problems, and training and coaching needs. Effective monitoring programs are closely aligned with both individual coaching and overall quality improvement initiatives.

There are several ways to monitor agents' performance; i.e., silent monitoring, call recording, side-by-side monitoring, peer monitoring, and mystery shoppers. Monitoring is also called position monitoring, quality monitoring or service observing.

M

Related terms: Calibration, Call Quality, Monitoring System.

Monitoring System. A system that records calls in order to have a permanent record of the complete transaction and to improve the quality of call handling.

Most-Idle Agent. Also known as the longest-available agent. A method of distributing calls to the agent who has been idle the longest. Related terms: Least-Occupied Agent, Next-Available Agent. (Vanguard)

Moves, Adds and Changes (MAC). As the term implies, MACs (pronounced "macks") are changes and/or additions to a telephone or data system.

Multilingual Agents. Agents who are fluent in more than one language.

Multimedia Routing and Queuing. Systems and processes that handle contacts across media—including voice, text-based and Web transactions—based on business rules that define how any transaction, inquiry or problem is processed. The key differentiator is not the media, but the customer and his or her need. Business rules should be established in regard to service level and/or response time for each alternative media.

Murphy's Law. The principle of pessimists that says, if anything can go wrong, it will. Not a good perspective to live by, but worth considering when designing agent groups, routing configurations and disaster recovery plans.

Music on Hold. Background music that callers hear when they are in queue or put on hold. See American Society of Composers, Authors and Publishers.

Mute. A telephone or headset feature that enables the user to deactivate the microphone (e.g., so that he or she can cough or carry on a side conversation without the other party hearing).

Mystery Shopper. A type of monitoring in which a person acts as a customer, initiates a call to the center and monitors the skills of the agent. See Monitoring.

Natural Language. Technology used in speech or
text recognition that identifies what is being said or
requested through free-form communication. No
structure or specific words or phrases are required.
See Speech Recognition. Related term: Directed
Dialog. (Vanguard)

Needs Analysis. In a training context, a systematic
and comprehensive process of assessing what training is needed in an organization. In a technology
context, a needs analysis defines the technology
solutions that are required to meet the organization's objectives, and how those solutions should be
configured.

Net Rep. A call center agent trained to handle
Internet transactions such as email, text-chat, Web
callbacks, co-browsing, etc. See Agent.

Netspeak. Abbreviated spelling and colloquial
phrasing employed by experienced Internet users.
For example, "BTW" for "by the way" and "IMHO"
for "in my humble opinion."

Network. In the call center world, the term network
is typically used to describe the inter-exchange
(IXC) services that route calls into a center or
among several centers. The network is the "pipe"
between the caller and the call center, or between

call centers. Related terms: Call-by-Call Routing, Network Control Center, Network Interflow, Percent Allocation.

Network Control Center (NCC). Also called traffic control center. In a networked call center environment, where people and equipment monitor real-time conditions across sites, change routing thresholds as necessary, and coordinate events that will impact base staffing levels. Related terms: Network, Network Management System.

Network Interface Card (NIC). A board inserted into a computer system that enables the system to be connected to a network. Most NICs are designed for a particular type of network and protocol, although some can accommodate multiple networks.

Network Interflow. A technology used in multisite call center environments to create a more efficient distribution of calls between sites. Through integration of sites using network circuits (such as T1 circuits) and ACD software, calls routed to one site may be queued simultaneously for agent groups in remote sites. Related terms: Call-by-Call Routing, Network, Percent Allocation.

Network Reports. Reports that provide information

on network call activity (e.g., network traffic, busies and call destinations). See Network.

Network Routing. The ability to make routing decisions in the network before selecting a location to route the call. (Vanguard)

Next-Available Agent. A call distribution method that sends calls to the next agent who becomes available. The method seeks to maintain an equal load across skill groups or services. When there is no queue, next-available agent reverts to longest-available agent. Related terms: Least-Occupied Agent, Longest-Available Agent.

Noise-Canceling Headset. Headsets equipped with technology that reduces background noise. See Headset.

Non ACD In Calls. Inbound calls that are directed to an agent's extension rather than to a general group. These may be personal calls or calls from customers who dial the agents' extension numbers.

Normalized Calls Per Agent. See True Calls Per Agent.

Number Portability. A shared database among network providers that enables call centers to keep the same telephone numbers even if they change carriers.

Occupancy. Also referred to as agent utilization or percent utilization. The percentage of time agents handle calls vs. wait for calls to arrive; the inverse of occupancy is idle time. For a half-hour, the typical calculation is: (Call volume x average handling time in seconds) / (number of agents x 1,800 seconds). The terms adherence to schedule and occupancy are often incorrectly used interchangeably. They not only mean different things, they move in opposite directions. When adherence to schedule improves (goes up), occupancy goes down. Further, adherence to schedule is within the control of individuals, whereas occupancy is determined by the laws of nature, which are outside of an individual's control. Related terms: Adherence to Schedule, Idle Time, Queue Dynamics, True Calls Per Agent.

Off-Peak. Periods of time other than the call center's busiest periods. Also a term to describe periods of time when long-distance carriers provide lower rates.

Off-the-Shelf. Hardware or software programs that are commercially available and ready for use "as is." Also refers to training programs that do not require customization.

Offered Call. Offered calls include all of the

O

attempts callers make to reach the call center. There are three possibilities for offered calls: 1) They can get busy signals; 2) they can be answered by the system, but hang up before reaching an agent; or 3) they can be answered by an agent. Offered call reports in ACDs usually refer only to the calls that the ACD receives. Related terms: Answered Call, Handled Call, Received Call.

On-the-Job Training (OJT). A method of training that exposes the employee to realistic job situations through observation, guided practice and while working on the job.

Open Ticket. A customer contact (transaction) that has not been completed or resolved (closed). Related terms: First-Call Resolution, Response Time.

Optical Character Recognition (OCR). Technology that reads printed text and determines what it says. Can be used with an imaging system to determine information about mail or fax items for routing and handling. (Vanguard)

Outsourcing. Contracting some or all call center services and/or technology to an outside company. The company is generally referred to as an out-sourcer or service bureau.

Overflow. Calls that flow from one group or site to another. More specifically, intraflow happens when calls flow between agent groups and interflow is when calls flow out of the ACD to another site.

Overlay. See Rostered Staff Factor.

Overstaffing. A scheduling term that refers to situations when the call center has more staff than is required to handle the workload.

Overtime. Time beyond an established limit (e.g., working hours in addition to those of a regular schedule or full work week).

Pareto Chart. Created by economist Vilfredo Pareto, a Pareto chart is simply a bar chart that ranks events in order of importance or frequency.

Payback Period. A capital budgeting method that calculates the length of time required to recover an initial investment.

PBX/ACD. A private branch exchange (PBX) that is equipped with ACD functionality. See Private Branch Exchange and Automatic Call Distributor.

Peaked Call Arrival. A surge of traffic beyond random variation. It is a spike within a short period of time. There are two types of peaked traffic—the type you can plan for, and incidents that are impossible to predict. Related terms: Increment, Traffic Arrival.

Peer Monitoring. Call center agents monitor peers' calls and provide feedback on their performance. See Monitoring.

Percent Allocation. A call routing strategy sometimes used in multisite call center environments. Calls received in the network are allocated across sites based on user-defined percentages. Related terms: Call-by-Call Routing, Network, Network Interflow.

Percent Utilization. See Occupancy.

Performance Driver. A suspect performance driver

that has been validated through statistically sound analysis. See Suspect Performance Driver.

Performance Objective. Usually stated as a quantifiable goal that must be accomplished within a given set of constraints, a specified period of time, or by a given date (e.g., reduce turnover by 20 percent within one year). Generally identified with fixing a problem or pursing an opportunity. See Goal.

Performance Target. An interim improvement point at a specific point in time, when striving to attain a new level of performance. The performance target is a "checkpoint" to assess progress and correct the action or work plans necessary to reach the final goal. Related terms: Key Performance Indicator, Performance Objective.

Peripheral Equipment. Equipment that is not integral to, but works with, a telephone or computer system (e.g., a printer or recording device).

Permanent Placement. A staffing term, related to using a staffing agency for the hiring process. The staffing agency handles all advertising and publicity and screens candidates using basic criteria (e.g., phone screen, testing). They may also handle other administrative tasks for the contracting party (e.g., reference or background checks, security clearance process). Typically, the call center handles face-to-face interviews and hiring decisions. A one-time fee is paid to the staffing agency, either at the time of hire or after a waiting period.

Personal Computer (PC). A computer designed for use by one person at a time. IBM compatible PCs and Apple Macintoshes are common examples of personal computers.

Personal Digital Assistant (PDA). A small, lightweight "palmtop" computer often used for personal organization tasks (e.g., calendar, database, calculator and note-taking functions) and communications (e.g., email, wireless Internet access and, in some cases, wireless telephone). Most PDAs use flash memory instead of disk drives, and rely on pens or other pointing devices rather than a keyboard or mouse.

Pilot Program. An experimental program to assess viability for a project or venture.

Poisson. A formula sometimes used for calculating trunks. Assumes that if callers get busy signals, they keep trying until they successfully get through. Since some callers won't keep retrying, Poisson can overestimate trunks required. Related terms: Erlang B, Retrial Tables, Trunk Load.

Pooling Principle. The powerful pooling principle

states: Any movement in the direction of consolidation of resources will result in improved traffic-carrying efficiency. Conversely, any movement away from consolidation of resources will result in reduced traffic-carrying efficiency. A common call center application is that if you take several small, specialized agent groups, effectively cross train them and put them into a single group, you'll have a more efficient environment (assuming all other things are equal). Related terms: Agent Group, Queue Dynamics, Skills-Based Routing.

Position Monitoring. See Monitoring.

Post-Call Processing (PCP). See After-Call Work.

Predictive Dialer. See Dialer.

Preview Dialer. See Dialer.

Primary Rate Interface (PRI). One of two levels of ISDN service. In North America, PRI typically provides 23 bearer channels for voice and data and one channel for signaling information (commonly expressed as 23B+D). In Europe, PRI typically provides 30 bearer lines (30B+D). Related terms: Basic Rate Interface, Integrated Services Digital Network.

Priority Queuing Application. Programming that recognizes and "bumps" higher-value customers up in the queue to ensure that they receive the most efficient service possible.

Privacy. The expectation that confidential or personal information disclosed in a private setting will not be made public. Call center managers should develop policies that address privacy-related issues. Important aspects of this responsibility include: 1) Communicate your policy regarding privacy; and 2) make employees aware of any electronic surveillance devices that are being used.

Private Automatic Branch Exchange (PABX). See Private Branch Exchange.

Private Branch Exchange (PBX). Also called private automatic branch exchange (PABX). A telephone system located at the call center's site that handles incoming and outgoing calls. ACD software can provide PBXs with ACD functionality. Many refer to a PBX as a "switch."

Private Network. A network made up of circuits for the exclusive use of an organization or group of affiliated organizations. Can be regional, national or international in scope and are common in large organizations.

Process. A system of causes. See System of Causes.

P

Products or Services Per Customer. A performance measure. A simpler variation of sales per customer, products or services per customer can be a measure of cross-selling effectiveness. In general, increases in the average number of products or services per customer are desirable and should increase customer value. Related term: Sales Per Customer.

Profit Center. An accounting term that refers to a department or function in the organization that generates profit. While call centers that are considered profit centers keep an eye on expenses, they also track value activities in the call center. Related term: Cost Center.

Prompted Digits. See Caller-Entered Digits.

Psychographics. The use of demographics to study and measure customer lifestyles, opinions, and preferences. Used for marketing purposes.

Public-Switched Network (PSN). See Public-Switched Telephone Network.

Qualitative Analysis. Analysis that interprets descriptive data, and is usually expressed as text. Related term: Quantitative Analysis.

Quantitative Analysis. Analysis that focuses on numerical, mathematical or statistical data. Related term: Qualitative Analysis.

Quantitative Forecasting. Using statistical techniques to forecast future events. The major categories of quantitative forecasting include time series and explanatory approaches. Time-series techniques use past trends to forecast future events. Explanatory techniques attempt to reveal linkages between two or more variables. Related terms: Forecasting Methodologies, Judgmental Forecasting.

Queue. Queue literally means "line of waiting people." (Note: Queue can also mean agent group—see Agent Group.) Queues are a fact of life in most incoming call centers because answering every call immediately would require as many agents as callers who need service at any given time. That is impractical for most organizations (although most emergency services centers do staff at levels that enable immediate answer much of the time). Related terms: Caller Tolerance, Invisible Queue, Visible Queue.

Q

Queue Display. See Readerboard.

Queue Dynamics. Queue dynamics refer to how queues behave; e.g., when service level goes up, occupancy goes down. Related terms: Agent Group, Average Speed of Answer, Occupancy, Service Level, Trunk Load.

Queue Time. See Delay.

Quick Disconnect. A modular connection that enables a headset to quickly be disconnected from a telephone set. See Headset.

Random Availability. The normal, random variation in the availability of agents due to variations in talk time, after-call work and call arrival. Related terms: Occupancy, Queue Dynamics.

Random Call Arrival. The normal, random variation in how incoming calls arrive. See Traffic Arrival.

Readerboard. Also called display board, queue display, wallboard or electronic display. A visual display, usually mounted on the wall or ceiling of a call center that provides real-time and historical information on queue conditions, agent status and call center performance. It can also display user-entered messages (e.g., "Happy Birthday, Grace!").

Real-Time Adherence Software. A function of workforce management software that tracks how closely agents conform to their schedules. See Adherence to Schedule.

Real-Time Management. Making adjustments to staffing and thresholds in the systems and network in response to current queue conditions. Related terms: Chaos Mentality, Queue Dynamics, Real-Time Report, Service Level.

Real-Time Report. Information on current conditions. Some real-time information is real-time in the strictest sense (e.g., calls in queue and current longest wait). Other real-time reports require some history (i.e., the last x calls or x minutes) in order to make a calculation (e.g., service level and average speed of answer). Related terms: Historical Report, Screen Refresh.

Real-Time Threshold. A marker that is identified in advance (e.g., number of calls in queue, longest in queue, etc.) that automatically initiates a certain response in a call center. For example, at a given time, a call center may not react to a queue unless it reaches 25 calls or more.

Received Call. A call detected and seized by a trunk. Received calls will either abandon or be answered by an agent. Related terms: Answered Call, Handled Call, Offered Call.

Recorded Announcement. A general reference to announcements callers hear while waiting in queue. Recorded announcements may remind callers to have certain information ready for the call, include general information about products or services, or provide alternative contact alternatives (e.g., "Visit our Web site at..."), etc. See Delay Announcement.

Recorded Announcement Device. A system component that enables recording, playback and control of delay and informational announcements, as well as music and/or other content played while callers are in queue or on hold. Related terms: Recorded Announcement, Delay Announcement.

Recorded Announcement Route (RAN). See Delay Announcement.

Redundancy. When a component or system is backed up by or shares the load with another component or system. Redundant systems ensure that, if there is a component failure, another component or system will keep the system going. See Disaster Recovery Plan.

Reengineering. A term popularized by management consultant Michael Hammer. Refers to fundamentally redesigning processes to improve efficiency and service.

Regional Bell Operating Company (RBOC). The individual companies that were created when the old Bell System was broken up into seven "baby bells." Only a few remain, due to mergers and acquisitions.

Remote Office. A group of agents (or a single agent working at a site separate from the main call center) that use the same technology infrastructure (voice switch, applications and other elements) as the main call center. The agent(s) could be located in a branch office or another center. Related term:

R

Satellite Office. (Vanguard)

Request for Information (RFI). A document sent to potential solutions providers that describes project requirements in high-level, generic terms. RFIs are generally issued to a broad range of possible vendors in order to become aware of the breadth and scope of possible solutions. See Request for Proposal.

Request for Proposal (RFP). A document sent to potential solutions providers that describes project requirements in focused, specific terms. RFPs are generally issued to fewer vendors than RFIs, and more specifically outline important criteria for potential solutions. See Request for Information.

Request for Quote (RFQ). An RFQ can be part of an RFI or RFP, or issued separately, and results in a price for a solution. Related terms: Request for Information, Request for Proposal.

Resolution. A measure of when a problem or issue is actually resolved. Used in environments where the call center's initial response may not fully resolve an issue. See Response Time.

Response Time. Defined as "100 percent of contacts handled within N days/hours/minutes" (e.g., all email will be handled within 240 minutes, or all faxes will be responded to within 24 hours). It is the preferred objective for contacts that do not have to be handled when they arrive. (See Service Level.)

Retention. The opposite of turnover; keeping employees in the call center. See Turnover.

Retrial. Also called redial. When a person tries again to complete a call after encountering a busy signal.

Retrial Tables. Sometimes used to calculate trunks and other system resources required. They assume that some callers will make additional attempts to reach the call center if they get busy signals. Related terms: Erlang B, Poisson.

Return on Assets (ROA). A ratio that divides net income (or earnings) by average total assets. The resulting percentage indicates how much income has been generated from each dollar of the organization's assets.

Return on Investment (ROI). Strictly speaking, this is the net income divided by total assets. In call center use, ROI has come to define a generic method of estimating the value of an investment. In this manner, the ROI of a project is typically calculated as the percentage return over the first year of the investment.

R

Return on Sales (ROS). A calculation that divides net income by sales to indicate if the return on sales is high enough. A low return on sales could indicate insufficient price mark-up to cover expenses.

Revenue. In a call center context, revenue often refers to a measure of the revenues attributed to call center services. Related term: Incremental Revenue Analysis.

Revenue Per Customer. See Sales Per Customer.

Ring Delay. Also called delay before answer. An ACD feature that enables the system to adjust the number of rings before the system automatically answers a call.

Root Cause. A primary cause of a problem or outcome. See System of Causes.

Rostered Staff Factor (RSF). Alternatively called overlay, shrink factor or shrinkage. RSF is a numerical factor that leads to the minimum staff needed on schedule over and above base staff required to achieve your service level and response time objectives. It is calculated after base staffing is determined and before schedules are organized, and accounts for things like breaks, absenteeism and ongoing training. Related term: Base Staff.

Round-Robin Distribution. A method of distributing calls to agents according to a predetermined list. Related terms: Next-Available Agent, Least-Occupied Agent, Longest-Waiting Agent.

Sales Force Automation (SFA). The use of computer and communications systems to support and boost the productivity of salespeople.

Sales Per Customer. A performance measure that can also be referred to as revenue per customer. Related term: Products or Services Per Customer.

Satellite Office. A call center location that operates using a cabinet or carrier of a switch from a main location. Used to extend one switch to another site to operate virtually without purchasing a second switch. Related term: Remote Office. (Vanguard)

Scatter Diagram. A quality tool that assesses the strength of the relationship between two variables. Is used to test and document possible cause-and-effect scenarios. See System of Causes.

Schedule. A plan that specifies when employees will be on duty, and which may indicate specific activities that they are to handle at specific times. A schedule includes the days worked, start times and stop times, breaks, paid and unpaid status, etc. See Schedule Alternatives.

Schedule Alternatives. Many scheduling alternatives exist that can help the call center efficiently meet staffing requirements that fluctuate throughout the day, week, month and year. Examples include: utilizing conventional shifts; staggering shifts; adjusting breaks, lunch, meeting and training schedules; forecasting and planning for regular collateral work; scheduling part-timers; establishing "internal" part-timers; offering concentrated shifts; offering overtime; giving agents the option to go home without pay; sending calls to a service bureau, and others.

Schedule Compliance. See Adherence to Schedule.

Schedule Exception. An activity not planned in an employee's schedule that becomes an "exception" to the plan. Related terms: Adherence to Schedule, Schedule, Schedule Alternatives.

Schedule Horizon. How far in advance schedules are determined.

Schedule Preference. A description of the times and days that an employee prefers to work. Related terms: Schedule, Schedule Alternatives, Schedule Horizon.

Schedule Trade. When agents are allowed to trade or "swap" schedules.

Scheduled Callback. A specified time that the call center will call a customer, usually based on the customer's preferences.

Scheduled Staff vs. Actual. A performance measure

that is a comparison of the number of agents scheduled vs. the number actually in the center, involved in the activities specified by the schedule. This measure is independent of whether or not you actually have the staff necessary to achieve a targeted service level and/or response time. It is simply a comparison of how closely reality aligned with the schedules you established. The purpose of the objective is to understand and improve staff adherence and schedules. Related term: Adherence to Schedule.

Screen Monitoring. A system capability that enables a supervisor or manager to remotely monitor the activity on agents' computer terminals. See Monitoring.

Screen Pop. A CTI application that delivers an incoming call to an agent, along with the data screen pertaining to that call or caller. Callers' records are retrieved based on ANI or digits entered into the IVR. Also called coordinated voice/data delivery. See Computer Telephony Integration.

Screen Refresh. The rate at which real-time information is updated on a display (e.g., every five to 15 seconds). Screen refresh does not correlate with the timeframe used for real-time calculations.

Seated Agents. See Base Staff.

Self-Service System. A system that enables customers to access the information or services they need without interacting with an agent. Web capabilities that enable online sales or services and IVR applications that provide automated flight information or access to financial accounts are common examples of self-service systems.

Service Bureau. A service bureau, sometimes referred to as an outsourcer, is a company hired to handle some or all of another organization's contacts. See Outsourcing.

Service Level. Also called telephone service factor (TSF). Service level is defined specifically as: "X percent of contacts answered in Y seconds"; e.g., 90 percent answered in 20 seconds. Contacts that must be handled when they arrive require a service level objective, and those that can be handled at a later time require a response time objective. Related Terms: Response Time, Service Level Agreement, Service Level Calculations.

Service Level Agreement (SLA). An agreement—usually between a client organization and an outsourcer (although they increasingly exist between departments within an organization)—which defines performance objectives and expectations.

Service Observing. See Monitoring.

Shrink Factor. See Rostered Staff Factor.

Shrinkage. See Rostered Staff Factor.

Signaling System 7 (SS7). A method of signaling within the voice network that uses a separate packet-switched data network (common channel signaling) to communicate information about calls. (Vanguard)

Silent Monitoring. See Monitoring.

Site Selection. The process of choosing a call center location that best meets the needs of the organization.

Six Sigma. Originally developed by Motorola, Six Sigma is a highly disciplined process that focuses on developing and delivering near-perfect products and services. Sigma is a statistical term that measures process variation. See System of Causes.

Skill Group. See Agent Group.

Skill Path. Skill paths focus on the development of specific skills rather than the progression of positions through the call center and/or organization.

Skill paths can move laterally (e.g., a printer technical support agent can be cross-trained to handle technical support on fax machines, as well) or upward (e.g., an agent can acquire leadership and coaching skills to add peer coaching responsibilities to his or her current position). See Career Path.

Skills-Based Routing (SBR). Skills-based routing matches a caller's specific needs with an agent who has the skills to handle that call on a real-time basis. SBR requires that the ACD be programmed with two "maps"; one map specifies the types of calls to be handled (e.g., gold-level, Spanish-speaking) and the other identifies the skills available by agent (e.g., Maria is capable of handling platinum, gold and silver calls, and she speaks English and Spanish). Related terms: Agent Group, Computer Simulation, Customer Access Strategy, Erlang C, Pooling Principle.

Smooth Call Arrival. Calls that arrive evenly across a period of time. Virtually non-existent in incoming call center environments. See Traffic Arrival.

Softphone. The ability to access telephony functions through a PC desktop computer interface instead of a telephone. (Vanguard)

Span of Control. The number of individuals a manager supervises. A large span of control means that the manager supervises many people. A small span of control means he or she supervisors fewer people.

Speaker Verification. A method of verifying the identity of a caller by comparing his voice to a previously stored voiceprint. Also referred to as voice authentication. See Speech Recognition. (Vanguard)

Special Causes. Variation in a process caused by special or unusual circumstances. Related terms: Common Causes, Control Chart.

Speech Recognition. Speech recognition enables IVR systems to interact with databases using spoken language, rather than the telephone keypad. There are two major types of speech recognition used in call centers today: 1) directed dialogue or structured language, which is prompting that coaches the caller through the selections; and 2) natural language, which uses a more open-ended prompt, recognizing what the caller says without as much coaching. See Interactive Voice Response.

Split. See Agent Group.

Split Shifts. Shifts in which agents work a partial shift, take part of the day off, then return later to finish their shift. Related terms: Schedule, Schedule Alternatives.

S

Staff Sharing. A staff-sharing relationship is when two or more organizations (or different units of an organization) share a common pool of employees, typically to meet seasonal demands. Related terms: Schedule, Schedule Alternatives.

Staff-to-Supervisor Ratio. See Span of Control.

Staggered Shifts. Shifts that begin and end at different times. For example, one shift begins at 7 a.m., the next at 7:30 a.m., the next at 8 a.m., until the center is fully staffed for the busy midmorning traffic. Related terms: Schedule, Schedule Alternatives.

Standard. A quantifiable minimum level of performance; performance below or outside the standard is not acceptable.

Strategic Staffing Plan. A forecast of future staffing requirements—which includes quantity and qualifications—generally over a one- to three-year timeframe. The plan focuses on three major issues:

1. The number of staff required

2. Required staff qualifications and associated development plans

3. Feasible workforce mix and scheduling alternatives

Related terms: Full-Time Equivalent, Schedule Alternatives.

Super Agent. See Universal Agent.

Supervisor. The person who has frontline responsibility for a group of agents. Generally, supervisors are equipped with special telephones and computer terminals that enable them to monitor agent activities. Related terms: Job Role, Monitoring, Span of Control.

Supervisor Monitor. Computer monitors that enable supervisors to monitor the call handling statistics of their supervisory groups or teams.

Suspect Performance Driver. A performance driver that has been attributed to results, but with an unproven linkage. Most people will have their own mental lists of suspected performance drivers. See Performance Driver.

Swat Team. The term some companies use for a team of non call center employees that act as "reservists" to quickly be assigned to call handling duties if the call load soars. Related terms: Schedule, Schedule Alternatives.

System of Causes. The variables that are part of a process. A call center is a process or system of causes. Taking a larger view, the call center is part of a larger process, the organization. In a lesser view, each agent group in a call center is a system of

S

causes unto itself, as are individual agents in a group.

Systems Integrator. A consulting firm or systems supplier that coordinates the installation and setup of complex systems. For example, call center applications may need to be integrated with legacy systems, requiring complex programming and project management.

T1 Circuit. A high-speed digital circuit used for voice, data or video with a bandwidth of 1.544 megabits per second. T1 circuits offer the equivalent of 24 analog voice lines. The European equivalent is known as E1 and it offers the equivalent of 31 analog voice lines. (Vanguard)

Talk Time. Everything from "hello" to "goodbye" in a phone call. In other words, it's the time callers are connected with agents. Anything that happens during talk time, such as placing customers on hold to confer with supervisors, should be included in this measurement. Also called direct call processing. Related terms: After-Call Work, Call Load.

Technology-Based Training (TBT). Training that uses technology to deliver instruction, typically outside of the formal classroom environment. TBT can include the use of computers or other technologies such as video or audiotape. See Computer-Based Training.

Telecommuting. Using telecommunications to work from home or other locations instead of at the organization's premises. Related terms: Remote Agent, Scheduling Alternatives.

Telemarketing. Generally refers to outbound calls for the purpose of selling products or services, or placing informational calls to customers, prospective customers or constituents.

Telephone Sales or Service Representative (TSR). See Agent.

Telephone Service Factor (TSF). See Service Level.

Temporary Employee. An employee hired for short-term projects or seasonal workloads. Temporary employees, often called "temps," are typically a good fit for assignments that last six to nine months or less.

Temporary-to-Permanent Placement. With this arrangement, job candidates are initially hired as an employee of the staffing agency and contracted on a temporary basis to the organization. See Temporary Employee.

Text-Chat Application. Allows customers visiting the corporate Web site to have real-time, text-based conversations with live agents. Text-chat applications also can provide agents with appropriate text templates to insert in their responses. In addition, they can enable agents to co-browse Web pages with customers and "push" specific pages to the customer. Agents can also be enabled to move the customer's cursor and help fill in complex forms and applications.

Text-to-Speech (TTS). Enables a voice processing system to speak the words in a text field using synthesized—not recorded—speech. Sometimes used for large, dynamic database applications where it is impractical to record all speech phrases, such as addresses or product names. Also used to "read" email or other text-based information over the telephone. (Vanguard)

Thick Client. A workstation in a client/server environment that performs much or most of the application processing. It requires programs and data to be installed on it and a significant part of the application processing takes place on the workstation. The client is "thick" in that it has much of the smarts of the overall application running on it. Related terms: Client, Thin Client. (Vanguard)

Thin Client. A workstation in a client/server environment that performs little or no application processing. Often used to describe browser-based desktops. The client is "thin" in that the applications reside on and are run within the server rather than the client. Related terms: Client, Thick Client. (Vanguard)

Threshold. The point at which an action, change or process takes place.

T

Tie Line. A private circuit that connects two ACDs or PBXs across a wide area.

Tiered Scheduling. An approach to allocating resources that defines a range of staffing requirements for a given time interval and places individuals in separate groups (tiers) within that range. For example, tier 1 may be scheduled for phone duty regardless of queuing conditions, but tier 3 won't sign on unless there are 15 or more calls in queue. Related terms: Schedule, Schedule Alternatives.

Time-Series Forecasting. See Forecasting Methodologies.

Toll-Free Service. Enables callers to reach a call center out of the local calling area without incurring charges.

Touchtone. A trademark of AT&T. See Dual-Tone Multifrequency.

Traffic Control Center. See Network Control Center.

Traffic Engineering. Designing telecommunications and data systems and networks to meet user requirements. Related terms: Computer Simulation, Erlang B, Erlang C, Traffic Arrival.

Training Evaluation. The process of assessing the impact of a training program. When assessing a training program, there are four levels of evaluation: Level 1: Reaction; Level 2: Learning evaluation; Level 3: Application to job; and Level 4: Evaluating the impact and ROI.

Training Strategy. A call center training strategy determines intermediate to long-term training priorities, objectives and direction.

Transaction. See Call.

Transmission Control Protocol/Internet Protocol (TCP/IP). A standard set of protocols that govern the exchange of data between computing systems. In call centers, TCP/IP is the underlying protocol of VoIP. It is also widely used in IVR, CTI and CRM systems. (Vanguard)

Trouble Ticket. The report of a customer's problem with a particular device or system, which is tracked through the workflow process. Trouble tickets were originally written on paper, but electronic trouble tickets are now standard in many workflow and help-desk applications.

True Calls Per Agent. Also called normalized calls per agent. It is actual calls (contacts) an individual or group handled divided by occupancy for that period of time. Related terms: Adherence to Schedule, Call Quality, Contacts Handled (Calls Per Agent).

T

Trunk. Also called a line, exchange line or circuit. A telephone circuit linking two switching systems. See Trunk Load.

Trunk Group. A collection of trunks associated with a single peripheral and usually used for a common purpose. Related terms: Trunk, Trunk Group.

Trunk Load. The load that trunks carry. Includes both delay and talk time.

Trunks Idle. The number of trunks in a trunk group that are non-busy.

Trunks in Service. The number of trunks in the trunk group that are functional.

Turnover. When a person leaves the call center. Turnover can be categorized as voluntary or involuntary. Voluntary turnover is when the employee decides to leave the organization or position. Involuntary turnover occurs when management makes the decision to end the employment relationship.

Unavailable Work State. An agent work state used to identify a mode not associated with handling telephone calls.

Unified Reporting. When data from different channels and systems are included on one reporting tool. This supports better analysis and decision-making in the organization.

Uniform Call Distributor (UCD). A simple system that distributes calls to a group of agents and provides some reports. A UCD is not as sophisticated as an ACD, and UCDs usually use simple hunt groups for call distribution. See Hunt Group.

Uniform Resource Locator (URL). The address for a Web page that is translated to an IP address.

Universal Agent. Also known as super agent. Refers to either: A) an agent who can handle all types of incoming calls, or B) an agent who can handle all channels of contact (e.g., inbound calls, outbound calls, email, text-chat, etc.).

Upsell. A suggestive selling technique of offering more expensive products or services to current customers during the sales decision. Related terms: Customer Profiling, Customer Segmentation.

Upsell and Cross-Sell Ratios. The percentage of

U

attempts to upsell or cross-sell that are successful.

Variance Report. A report illustrating budget/cost objectives that look at the difference between projected and actual expenditures for various budget categories.

Video Display Terminal (VDT). Another term for computer monitor; a data terminal with a TV-like screen.

Virtual Call Center. Also called distributed call center. Multiple networked call centers that operate as a single logical system even though they are physically separated and geographically dispersed. This permits economies of scale in call-handling, as well as supporting disaster recovery, call overflow and extended hours of coverage. (Vanguard)

Virtual Private Network (VPN). A method for using a public network (like the Internet) for a company's private business purposes. To address security, information is encrypted before being sent and then decrypted at the receiving site. (Vanguard)

Visible Queue. When callers know how long the queue that they just entered is, and how fast it is moving (e.g., they hear a system announcement that relays the expected wait time). Related terms: Invisible Queue, Queue.

Voice Extensible Markup Language (VXML). An

emerging standard for developing voice-processing
(IVR) applications with Internet and Web-based
tools. The vision of VXML is that millions of Web
developers will be able to develop IVR and speech
recognition applications, based on a familiar programming format. (Vanguard)

Voice over Internet Protocol (VoIP). Transmitting
voice conversations as packets of data from one
communications device (voice switch, PC or IP
phone) to another over a TCP/IP network.
(Vanguard)

Voice Processing. An umbrella term that refers to
any combination of voice technologies, including
voicemail, automated attendant, audiotex, interactive voice response and faxback. See Interactive
Voice Response.

Wallboard. See Readerboard.

Wallet Share. Related to customer retention is wallet share, also called share of wallet. This refers to the amount of a customer's total spending in a product category that goes to your organization.

Web Call. A voice call initiated by a customer from a company's Web site. Web calls can be accomplished in two ways: the caller can speak by VoIP over the Internet or be immediately called back over the PSTN. (Vanguard)

Web Call-Through. Using voice over Internet (VoIP) technology, the customer clicks on a button that establishes a voice line directly to the call center.

Web Callback. By clicking on a button, the customer lets the company know that he/she wants to be called back either immediately or at a designated time.

Web Collaboration. A broad term referring to the ability for an agent and customer to share content by pushing/pulling Web pages and/or whiteboarding and page markup.

Web Integration. Incorporating Web contact into the call center by providing access to an agent over the Internet when needed. Can be enabled through text-chat or a Web call. Email is sometimes offered as part of this integration. Often includes "co-browsing" or "pushing" Web pages to the customer. (Vanguard)

Web Self-Service Tools. Tools that enable customers to receive information and answers to questions, place orders and view order status directly from the corporate Web site without contacting the call center for assistance. See Self Service System.

Whisper Transfer. An IVR integration technique where the IVR temporarily connects to the agent and speaks the account number or other information before connecting the caller to the agent. (Vanguard)

Wide Area Network (WAN). The connection of multiple geographically dispersed computers or LANs, normally using digital circuits. The device that connects A LAN to a WAN is usually a router. Related term: Local Area Network. (Vanguard)

Wide Area Telecommunications Service (WATS). WATS has become a generic term for discounted toll services provided by long-distance and local telephone companies.

Wireless Application Protocol (WAP). A carrier-independent protocol for wireless networks, designed to enable wireless users to access a new generation of multimedia and Web-based services.

Work State. An ACD-produced indicator of the status of a call center agent's activity or status. See Agent Status.

Workflow. A business application that enables work tasks to be executed consistently and thoroughly, driven by business rules. The movement of each task can be tracked throughout the duration of the process, providing both current status and historical activity. (Vanguard)

Workforce Management System (WFMS). Software systems that, depending on available modules, forecast call load, calculate staff requirements, organize schedules and track real-time performance of individuals and groups. Workforce management can be performed for a single site or for networked sites. In a multisite environment, forecasting and scheduling may be performed at a central site or in a decentralized fashion at each site. Tracking and adherence monitoring is generally a local function. Related terms: Computer Simulation, Erlang B, Erlang C, Forecasting Methodologies, Queue Dynamics.

Workload. Often used interchangeably with call

load. Workload can also refer to non-call activities. See Call Load.

World Wide Web (WWW). The capability that enables users to access information on the Internet in a graphical environment.

Wrap-Up. See After-Call Work.

Wrap-Up Codes. Codes that agents enter on their phones to identify the types of calls they are handling. The ACD can then generate reports on call types by handling time, time of day, etc. Wrap-up codes are generally entered at the completion of each contact, although some systems enable agents to enter wrap-up codes while in talk time.

Zip Tone. See Beep Tone.

Sources

The following sources were used in this project:

INCOMING CALLS MANAGEMENT INSTITUTE (ICMI)

410-267-0700
www.icmi.com
icmi@icmi.com

TARP

703-524-1456 (U.S. number)
www.tarp.com
info@tarp.com

VANGUARD COMMUNICATIONS CORP.

973-605-8000 (U.S. number)
www.vanguard.net
info@vanguard.net

How to Contact the Author

Do you have suggestions for future editions?
Comments? Feedback? Please contact us!

Incoming Calls Management Institute (ICMI)
410-267-0700
www.icmi.com
icmi@icmi.com
Brad Cleveland, bradc@icmi.com

About ICMI

ICMI Inc. is a global leader in call center consulting, training, publications and membership services. ICMI's mission is to help call centers (contact centers, help desks, customer care, support centers) achieve operational excellence and superior business results.

Based in Annapolis, Maryland, the organization was established in 1985 and was first to develop and deliver management training customized for call centers. Today, ICMI has become the industry's leading provider of membership services with an impressive line-up of call center management resources, including instant access to prominent research, expert advice and career development tools, and a networking forum that spans more than 40 countries worldwide.

ICMI is not associated with, owned or subsidized by any industry supplier—its only source of funding is from those who use its services. For more information about ICMI, visit www.icmi.com, or call 800-672-6177 (410-267-0700).

Order Form

Item	Member Price	Price
Call Center Management On Fast Forward: Succeeding In Today's Dynamic Inbound Environment**	**$23.76**	$34.95
Call Center Technology Demystified: The No-Nonsense Guide to Bridging Customer Contact Technology, Operations and Strategy**	**$33.96**	$39.95
ICMI's Call Center Management Dictionary: The Essential Reference for Contact Center, Help Desk and Customer Care Professionals**	**$21.21**	$24.95
ICMI's Pocket Guide to Call Center Management Terms*	**$5.12**	$5.95
ICMI Handbook and Study Guide Series Module 1: People Management*** Module 2: Operations Management*** Module 3: Customer Relationship Management*** Module 4: Leadership and Business Management***	**$169.15 each**	$199.00 each

For more information on our products or to order, please visit **www.icmi.com**

Order Form

Item	Member Price	Price

Topical Books:
The Best of *Call Center Management Review*
  Call Center Recruiting and New Hire Training*
  Call Center Forecasting and Scheduling*
  Call Center Agent Motivation and Compensation*
  Call Center Agent Retention and Turnover*
	$14.41 each	$16.95 each

Forms Books
  Call Center Sample Monitoring Forms**
  Call Center Sample Customer Satisfaction Forms Book**
	$42.46 each	$49.95 each

Software
  QueueView: A Staffing Calculator—CD ROM*
  Easy Start™ Call Center Scheduler Software—CD-ROM*
	$41.65
	
	$254.15
		$49.95
		
		$299.00

Call Center Humor: The Best of *Call Center Management Review* Volume 3*	**$8.45**	$9.95
The Call Centertainment Book*	**$7.61**	$8.95

Notes

Notes